Wireless Etiquette

Wireless Etiquette

A Guide to the Changing World of Instant Communication

Peter Laufer

OMNIPOINT COMMUNICATIONS
New York

Library of Congress Cataloging-in-Publication Data

Laufer, Peter
  Wireless Etiquette, A Guide to the Changing World
  of Instant Communication / by Peter Laufer
p. cm.
ISBN 1-892918-00-5   98-96554

United States Constitution, First Amendment
Congress shall make no law respecting an establishment of
religion, or prohibiting the free exercise thereof; or abridging
the freedom of speech, or of the press; or the right of the people peaceably to assemble, and to petition the Government
for a redress of grievances.

Wireless Etiquette is available at special discounts for bulk pur-
chases, sales promotions, fund-raising, or educational purpos-
es. Special editions can be created to specifications. For
details, contact Omnipoint Communications.

First published in 1999 by
Omnipoint Communications
16 Wing Drive
Cedar Knolls, NJ 07927

Every man, woman, and child

should carefully examine the workings of Prof. Bell's

speaking and singing telephone, in its

practical work of conveying instantaneous

communication by direct sound, giving the tones of

the voice so that the person speaking can be recognized

by the sound at the other end of the line.

FROM AN 1877 PLACARD ADVERTISING "AN
EXHIBITION WHERE ALL VISITORS DESIRING
CAN MAKE FOR THEMSELVES A PRACTICAL
INVESTIGATION OF THE TELEPHONE."

If you answer that phone, I'm sending you to jail.

ALLEGHENY COUNTY (PENNSYLVANIA)
JUDGE DAVID CASHMAN TO A LAWYER
REACHING FOR A RINGING MOBILE TELEPHONE
DURING A TRIAL

CONTENTS

I'll admit it: I love wireless telephones. I also hate them. Let me tell you why.

As a CBS News correspondent, I found myself witness to an armed uprising in the Russian capital during the first week of October, 1993.

Hundreds of military troops surrounded the Parliament building in central Moscow while anti-Yeltsin forces holed up inside, seeking to bring down the reform government.

As the conflict escalated, the Kremlin cut off power and telephone lines, leaving its opponents in the dark—literally and figuratively.

Wireless phones can be an invaluable alternative to their conventional counterparts.

Along with dozens of colleagues from the international press corps, I cajoled my way across the front line of this emerging battle zone to cover the controversy.

Without any normal means of communication inside the Russian White House, I relied on a clandestine portable wireless telephone to report the news. It was the best way for me to keep in touch. It was also a way for the rebels to get their message to the outside world—something Yeltsin's men didn't want. The standoff continued for several tense days. I crossed back and forth from one armed camp to the other, portable phone in tow, reporting what looked like the makings of all-out civil war.

Eventually, my cell phone was banned from the building. Finally, so was I. Hours later, the shelling started and I continued my coverage from outside—via cell phone.

I couldn't have done my job without a mobile unit back then. But it also caused me to be temporarily detained and nearly jailed by federal security agents.

Of course, these days, I can't imagine living or working without wireless telephones–especially since my employer, Omnipoint Communications, operates one of the biggest wireless networks in the United States.

However, I also recognize the potential for their misuse. The spread of this advanced technology enables personal communication devices to follow us everywhere–from home and office to customary places of peace and privacy. We are faced with the challenge of previously unimagined contacts.

Most of us are genetically incapable of ignoring a ringing telephone. It doesn't seem to matter that modern technology also gives us answering machines and voice mail. Every day, living, breathing human beings get ignored in favor of unknown callers. Even the more recent caller ID feature–which lets us know who is electronically knocking at our virtual door–doesn't stop some of us from giving distant visitors priority over actual, present conversations.

There are occasions when words aren't necessary.

Let me tell you two stories.

I was recently meeting with a co-worker in my office. During our conversation, his phone started to ring. As he moved to answer it, I stopped him.

"What are you doing?" I asked.

"I should get that," he said.

"Why?" I pressed.

"It might be someone important," he explained, gesturing urgently to the still-ringing phone.

"And that would make me…?" I left that thought unfinished, but I think he got my point.

One more anecdote. This one's apocryphal but good enough to repeat.

An unemployed engineer was making the rounds applying for work. She carried a wireless phone with her to be sure she wouldn't miss any job opportunities. Right in the middle of an interview, her mobile unit began to ring. A dilemma: should she answer it or shouldn't she? What if the unknown caller were offering a better job than the one she was now discussing? Clearly, this prospective employer would not like the interruption, but could she afford not to know?

(For the sake of a happy ending, we'll say that she got the job and the unanswered call turned out to be a wrong number. Or, she answered the call, but it was a radio station contest and she won a million dollars. You choose.)

As our gadgets get smarter, faster, and more complex, it seems we can never quite keep up with them. Some days, it's as if the game changes before we know what the rules are. Some days, it's as if there are no rules.

This book offers some new perspectives on the polite use of wireless phones. The same principles could just as easily apply to laptop computers, personal stereos, or any other portable modern technological tools. Whether *Wireless Etiquette* is for yourself or someone else, here's hoping it will lead to better communications for us all.

Terry Phillips
New York City, 1998

1
WHICH WAY

When the elevator was invented, people were forced to deal with the new experience of being jammed into a confined arena with total strangers in order to move efficiently from floor to floor in skyscrapers. Quickly social norms were established: gentlemen removed their hats for the ride, passengers faced forward toward the door, probably to avoid awkward face-to-face encounters, ladies were invited to enter and exit first. Back in the good old days hotels, office buildings, and apartment houses employed elevator operators who then became the arbiters of proper elevator etiquette. "Please face the front of the car," they would coo with some authority as the elevator filled. By the time the smartly uniformed operators were replaced with automatic machinery, habits had already been established. Most of us face elevator doors still and, if we're standing near the control panel, take over the role of the operator and politely ask new passengers what floor they seek.

When our streets and highways filled with automobiles, we were forced to accept rules and regulations enforcing good behavior—lives were at stake.

The availability of cheap, portable cassette and CD players introduced us to a new form of rudeness: the be-bopping passerby who blasts us with his or her favorite tunes and assumes we like them too, doesn't care if we're disturbed, or in fact enjoys the rebellious public disturbance.

The list is long of inventions that influence behavior.

Good manners are never out of style

Society seeks guidance for proper conduct. Long before Miss Manners started her etiquette newspaper column, the books of Emily Post and other architects of style shepherded Americans through changing social mores. Abigail Van Buren and her twin sister Ann Landers do their part in shaping manners and customs, as do the more avant-garde advisors in alternative newspapers. Even academia gets involved. There is a course at Johns Hopkins University called "Civility, Manners, and Politeness," and the *Wall Street Journal* found a course syllabus at Utah State that includes these specific instructions: "This class is too large for chitchat; please do not." The orders explain "how disturbing it is to your fellow classmates to be forced to endure your idle chatter and giggling. The students who sit near you are not interested in your romantic lives, how out of touch you think your parents are, how stupid you think your teachers are, etc."

P.C. Vey

"He's not in. Can I take a message?"

The voice with a smile

So it was when the telephone was invented. How to use this new device politely settled into what seemed practical and what felt appropriate. And self-appointed etiquette counselors such as Ms. Post offered their advice. Even the device's inventor Alexander Graham Bell injected himself into the fray by trying to influence conversationalists that "Hoy! Hoy!" instead of "Hello!" would be the best greeting when they answered the new invention's demanding ring.

Since those early days we've learned to deal with the various idiosyncrasies of telephone development.

Few telephone subscribers must contend with party lines these days, but just what was appropriate to do when you picked up the telephone and your neighbors were gossiping on the party line required some attention back when those lines were common. (If the gossip was any good, the most common practice undoubtedly was to cover the mouthpiece and listen.)

We were forced to come up with an acceptable response to an incessantly ringing phone during business meetings or formal dinner parties. Do we respond to the interruption—intrigued with the unknown it brings—or leave our attention with our in-person correspondents? The answering machine and caller ID helped solve that dilemma, offering us the opportunity to screen calls or save the incoming information for later. And we learned how to terminate phone conversations when other activities drew our attention.

One of the more difficult challenges for polite society created by the telephone continues without a clear solution: how to deal with call waiting. When the persistent beep tells you that another call is coming in to your number, you must either ignore it or inform your current correspondent that you intend to leave the line. But once you answer the second call you must make a difficult decision: which call is the more important to you. Once you decide, you must then inform the caller you are rejecting, an awkward task to accomplish without appearing rude, since your inherent message is clear. The other call and caller is more valuable to you than the one you disconnect.

Fashion model
James King

Judging from a letter his wife wrote shortly after his death, Bell himself would not have added call waiting to his phone service. "He never had one in his study," Mabel Bell wrote about her husband's invention. "That was where he went when he wanted to be alone with his thoughts and his work. The telephone, of course, means intrusion by the outside world."

You're a number and a name
Now the latest developments in wireless communications are added to the mix of technology etiquette. Today's mobile phones combine wireless voice and data connections to a single number that is assigned to each user, not to the handset. They offer spectacular opportunities to keep in touch with friends and family and do business

with virtually anyone from virtually anywhere, and their use is becoming ubiquitous in our social milieu. As such, opportunities for rude behavior abound. Mobile telephone use is changing our lifestyles and our environments.

The rapid growth of wireless technology is overtaking society's ability to accommodate the saturation of mobile phones in our midst without some common guidance. Because of the one-to-one relationship of caller to caller during a telephone conversation, many insensitive mobile phone users are rude without intention. They merely forget their physical surroundings because they become so completely caught up in their telephone conversation.

Bernard Schoenbaum

"Mr. Watson, come here. I want you," Alexander Graham Bell is reputed to have said as the first words ever spoken over a telephone. Phenomenal advances mark the relatively short history of the telephone, from Bell's famous cry for help through to the ability today to communicate virtually instantly from anywhere.

On March 3, 1876, Bell won patent number 174,465 from the U.S. Patent Office for the telephone (a word created from the Greek words for "far" and "sound"). Only a week later he and his assistant Thomas Watson connected two rooms in Bell's Boston laboratory with the first telephone wires, and—goes the legend—then Bell spilled some battery acid and made the first-ever telephone call. The line was scratchy and the sound transmitted was far from the true-to-life replication of human voice that we enjoy today. But Bell and Watson were obviously ecstatic. After years of dreams and experiments, the telephone worked.

Deafness and insanity

Just as is the case today with the spread of wireless technology, the early telephones were greeted with fear and suspicion. Bell and Watson tried to intrigue the public with the new invention by staging demonstrations of long-distance conversations between the Boston laboratory and lecture halls around the state. Some in the audience assumed the connections were just a trick. Others feared that the voices coming through a telephone receiver could cause deafness or insanity. Telegraph companies, worried about competition from the telephone, fueled the rumors.

But the telephone prevailed quickly. The year after Bell's famous call for help, an exchange was built in Boston to connect the growing number of telephones in use around the city. And the next year inventor Bell dreamed of the future as he wrote publicity for the new device: "It is conceivable that cables of telephone wires could be laid underground or suspended overhead, communicating by branch wires with private dwellings, uniting them through the main cable with a central office, establishing direct communications between any two places in the city. I believe that in the future wires will unite the head offices of the Telephone Company in different cities, and a man in one part of the country may communicate by word of mouth with another in a distant place."

Even Alexander Bell was not yet ready for wireless telephone connections.

But within the next ten years Americans were thrilled by the practicality and romance of using the telephone, and more than 150,000 phones were in service around the United States by 1887. In 1915 Bell and Watson talked with each other over the first transcontinental connection.

The first wireless
About twenty years after Bell successfully called Watson, Guglielmo Marconi broadcast voice through the air, and radio was born. The marriage of the technologies eventually created the mobile telephone.

Early radio was called wireless telegraphy, and shortly after Marconi's successful experiments it was widely used to communicate with ships at sea. During World War I, after Germany severed Great Britain's telegraph connections, Marconi built radio stations for the empire to allow it to communicate with its worldwide colonies. The same technology was employed on land by police departments, with some of the earliest radio cars, running on the streets of Detroit in the 1920s, receiving orders from headquarters. By the time of World

War II, radio transmitters small enough for a single soldier to carry were in use, including the first portable two-way radios.

Finally mobile telephones

After the Second World War, a system was devised to connect mobile wireless voice transmissions with the wired telephone network. The service was limited to calls manually connected between the two systems by an operator. Mobile callers picked up their handsets, pushed a transmit button, and called for the operator, providing the number to be called. It was the operator who actually placed the call. One tower and antenna were used to connect the mobile units to the wired phones, and in these early days of mobile phone service only a handful of channels was allocated to each metropolitan area. Mobile callers needed to wait for a clear channel by picking up the phone and listening to hear if a line was in use, making it easy to explain away eavesdropping. By the 1960s switches were developed that automatically searched for open channels. But even as late as the mid-1970s, service was still severely limited. The entire New York City system could only handle 543 calls at one time.

WWII era mobile phone next to contemporary cell phone

This cumbersome system was finally replaced with the cellular concept. The cellular phone system divides territory into cells, each with its own base station transmitter and receiver. Each mobile phone communicates with these base stations, and

*Motorola
DynaTAC
circa 1973*

as the phone user moves out of one cell's range an automatic switching system hands the call off to the next adjacent base station. Because the power which transmitters need to cover the relatively small area of a cell is much less than what was needed to cover an entire metropolitan region, frequencies can be reused by nearby cells without interference. As such, use of the radio spectrum allocated to cell phones is much more efficient than that of the early mobile phones.

Cellular telephones were introduced into commercial service in 1983. These were what are known as analog phones, transmitting voice over radio waves using the same basic technique the Detroit police cars used for their receivers back in the 1920s. Consumer demand for mobile telephones was immediate and far exceeded even the most ambitious industry and government projections. By the mid-1990s, what is called Second Generation systems were licensed and built, employing sophisticated digital technology. In addition, commercial mobile phones were developed that connect to the worldwide telephone system via satellites. These expensive units allow for the making of phone calls from the wilds of the Australian outback to the ice fields of Siberia, just about all of the world's most remote places.

Clear and reliable

The Second Generation digital wireless systems provide extraordinary sound reproduction. They are able to transmit complex data in addition to voice, a long distance from the noisy first telephone lines of Bell and his assistant Watson. Looking back on his career developing the telephone, Thomas Watson mused about the quality of those early circuits: "I used to spend hours at night in the laboratory listening to the many strange noises in the telephone and speculating as to their cause. One of the most common sounds was a snap, followed by a grating sound that lasted two or three seconds before it faded into silence, and another was like the chirping of birds. My theory was that the currents causing these sounds came from explosions on the sun or that they were signals from another planet. They were mystic enough to suggest the latter."

Contemporary digital wireless technology is still mystic for most of us, since the connections we can make through tiny handsets seem like such magic.

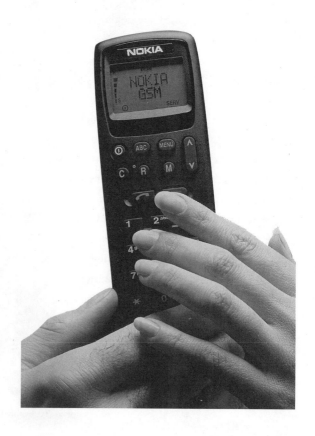

Compared with today's incredible pocket-sized wireless instruments, the mobile telephones of just a few years back look impossibly impractical.

The early radio telephones were carried in big backpacks. Even the first cellular phones of the early 1980s were cumbersome and heavy affairs, laden with batteries and sure to attract attention whenever used in public.

Snake oil?

One side effect of the questions surrounding health and mobile phone use is the marketing of products advertised as designed to neutralize possible danger from phones.

One is the Q-Link (for "quantum link") Pendant, which promises "enhanced, natural protection from computers, cell phones, and other sources of man-made electromagnetic fields." The Q-Link Pendant, claims its promoters, "is based on the emerging science of subtle energies, which has discovered that a pure source of subtle energy can boost human bioenergy."

*The Q-link
Pendant*

The Q-Link Pendant goes for $129 plus tax, shipping, and handling from Nutrinet, Inc., of Tustin, California.

Ads for another product, the Phone/Shield, promise that it "blocks radiation hot spots produced by your cellular phone." The device slips over the earpiece and antenna of the handset and, according to the marketer, "dramatically reduces the radiation directed at your head by reflecting it away." From Italy comes SCUDO, offered as "a complete protection system featuring special internal radiation blocking carbon fiber, reinforced with nickel and other materials." SCUDO is hand-stitched Italian leather, "a smart way to protect your brain and your phone at the same time," suggests a Ghent, New York, firm called Less EMF, Inc., which markets Phone/Shield for $29.95 and SCUDO for $39.95.

A new language

Mobile phones influence language, and now an alphabet soup of jargon accompanies the industry. Among the common new terms are: PCS for Personal Communications Services, GSM for Global System for Mobile communications, and SIM for Subscriber Identity Module (the little chip that identifies a GSM phone user's number). Nicknames for the phones themselves vary from culture to culture. The British refer to theirs as "mobiles." "Handy" is the term preferred in Germany. In Singapore, hand-held phones are called "prawns," because the first popular models

Alana Roth

Sign in Mexican airport

NO USAR A PARTIR DE ESTE PUNTO

DO NOT USE BEYOND THIS POINT

seen all over the city flipped open into a shape similar to a prawn. The French use *"portatif,"* a new and unexpected application of a quirky, antiquated word. In Italy the melodic diminutive *"telefonini"* is used, as in the sign at La Scala, in Milan, mandating etiquette: "Leave telefonini in the cloakroom."

*A grieving Israeli family
decided to honor their dead loved one's
passion for his mobile phone with this
final statement.*

4 THE WRONG WAY

Most wireless companies offer a convenient directory-assistance service that finds the number you're seeking and then proceeds to connect you for no additional charge. On the subject of the information service, Emily Post, in her first

Bill Woodman

"Fenwick, Benton & Perkins. How may I direct your call?"

post–World War II edition of her standard, *Etiquette*, showed how even she, with her established credentials as an etiquette expert, could be as rude as anyone else.

"…it is not the the feebly old or the weakly ill who call upon 'Information' to a degree that is literally crippling the service," she wrote. "Investigation has proved," she continued without documentation, "that the principal offenders are the lazy young who have seemingly neither the strength of muscle nor the sense of fairness to lift themselves off their spines long enough to look up a number in a city telephone book; they ask 'Information' to do it for them."

Nasty words.

And Ms. Post was quite the protector of the Telephone Company (she always made the generic a proper noun) at the time, suggesting in one section of her book that we should show "consideration" to the Company by "taking care of the instrument itself." Her specific advice: "Don't stand it where it may get easily knocked off, or too near a radiator, or in the very hot sun or where it can be rained on. Don't let the cord get snarled tight!" The exclamation point is definitely hers.

There's a right way and a wrong way

To accompany its new technology with social mores, telephone companies early on got involved in trying to influence phone manners, teaching customers how to use the equipment and how to act during conversations. "Hold the transmitter directly in front of the lips while you are talking," explained a 1940 Bell System advertisement. Other telephone etiquette guidelines from the ad pointed out that "thoughtlessly slamming the receiver may seem discourteous,… shouting distorts the voice and may make it gruff, and unpleasant,… A good impression is made by the sound of the voice over the telephone."

Rules of etiquette are looser these days, but just because there are new tools and toys available doesn't mean we need to add to the degeneration of decent manners just because we can. It is easy to become addicted to new concepts and technologies. People who just a few years ago scoffed at the ideas of voice mail and call waiting now cannot conceive of their personal or business lives being without these services. Similarly, wireless communications has moved from novelty to utility for a growing segment of the population. The understandable desire to take advantage of such technological opportunities is not in conflict with using the devices safely and with courtesy.

In fact, many of the features available with wireless phones help make it easy to use the phones politely.

Wireless communications now includes not just the simple telephone call but also can come equipped with answering machines, voice mail, e-mail, fax capability, pagers, call waiting, caller ID, call forwarding, and even Internet connections. News, weather, and sports reports can be ordered to appear as text on phone handset screens. The ringers can be turned off or switched to vibrate instead.

These features mean that your phone need not be a social interruption. You can silently read e-mail from the Internet without disturbing others. You can switch the handset to its answering machine mode and periodically collect your messages without saying a word. A page can come through to you as a subtle vibration in your pocket

"Hi, honey—I'm home."

—unknown to those around you—and you can read a brief page message on the screen of your phone. With the most sophisticated equipment, you can silently connect to the Internet and surf the Web with a tiny keyboard hidden inside the handset.

Considering the growing concern about the potential distractions of driving while talking on the telephone, voice-activated dialing is a valuable feature. Wireless devices can be programmed to recognize your voice when you say the name or number of people you wish to call.

A case against
cute kindergartners on answering machines

When you record the message for the answering machine feature of your wireless telephone, it's best to skip the jokes and gimmicks. Especially for frequent callers, waiting through long and complex messages is time-wasting and annoying.

Some answering systems allow the caller to go directly to recording mode, bypassing your outgoing message with the touch of a button and saving time and patience.

Hey! Is somebody else on this line?

Even the most sophisticated hackers agree that the latest digital generation of mobile telephony is not susceptible to eavesdropping. But earlier analog systems that are still in use can be tapped, as Prince Charles, Newt Gingrich, and plenty of cheating spouses learned to their surprise.

Texas congressman Bill Sarpalius was tape recorded back in 1990 trying to set up a date with a twenty-year-old woman he called "cute" and offered to help obtain a job. A journalism student at Amarillo College tried to sell the tape to radio and TV stations but was instead himself prosecuted and fined for intercepting the call.

A few years later House Speaker Newt Gingrich ignored this history and was taped off a radio scanner by a couple of Democrats in Florida, who shipped their copy of his remarks up to the Democratic leadership on Capitol Hill. Gingrich was embarrassed; he was heard trying to organize Republicans to oppose the decision that the Speaker be reprimanded for House ethics rules violations, lobbying he had promised he would avoid. Yet it was Democratic heads that rolled for leaking the transcript of the intercepted phone call to the *New York Times*, because such eavesdropping is illegal.

The most famous story of an intercepted cellphone call was the pillow talk between Prince

Charles and his lady friend Camilla Parker-Bowles.

"I can't bear a Sunday night without you," she was quoted around the world as saying on the lusty tape.

The future king of England was quoted as responding, "What about me? The trouble is, I need you several times a week."

If Speaker Gingrich and His Royal Highness had used digital phones, their private conversations would have stayed out of the headlines.

Haute couture

Courtesy of Motorola Museum
c. 1998 Motorola, Inc.

Marketers see a viable future in the mobile phone as fashion accessory. Philips marketing manager Joe Weller is convinced that his company's women customers choose phones by their color, their shape, and how the equipment "looks against their face." Nokia too targets its colorful phones toward women customers. At Ericsson, fashion concerns led to the model 638, featuring a replaceable front panel to allow users to match the color of their phones with their mood or clothing. Fifteen different hues

are available. Motorola's StarTAC handset is advertised as a "wearable" phone. The size of a pack of playing cards, it can be clipped onto a belt, slipped into a top pocket, or even hung on a cord like a necklace.

5 THE RIGHT WAY

In just one generation, mobile phones have established a secure place in popular culture lore, in part because they have such an impact on how we relate to each other socially.

There's no business like show business

There's a spectacular scene at the beginning of the comedy *For Richer or Poorer*, starring Tim Allen and Kirstie Alley, based on our ubiquitous use of hand-held telephones. The Tim Allen character

is being chased around Wall Street by IRS agents. He reaches into his inside jacket pocket for his phone to call for help. One of the agents figures it must be a gun and shoots the phone out of Allen's hand.

Tim Allen on the set

"I can't believe you just shot him!" complains the agent's partner.

"He had a gun!" yells the trigger-happy agent.

"He had a phone!" responds his shocked and disgusted partner.

But the mobile telephone had showed up even earlier as a movie character. In the original Audrey Hepburn version of the delightful romantic comedy *Sabrina*, her future lover, played by Humphrey Bogart, is commuting in an early Fifties limousine from Long Island to his Manhattan office. Bogart picks up a clunky looking black plastic telephone receiver from a bulky apparatus attached to the interior of the car.

"This is KO 7 5263," he tells an operator.

"Give me Bowling Green 9 1099."

The mobile phone in the movie is used as a device to express the Bogart character's affluence and importance.

"How did the market open?" he asks his secretary. With one of the current generation of mobile telephones he could read

Courtesy of Motorola Museum
c. 1998 Motorola, Inc.

specific stock prices on the screen of his handset without bothering his secretary. As he mulls over the market, he instructs the office, "I'm just leaving the house; you can put the coffee on in forty-five minutes."

The President's Analyst starring James Coburn is a funny period piece from 1967 that paints The Phone Company (TPC) as ultimate villain trying to control the world. As the picture builds to its climax, Coburn's character, psychiatrist to the president of the United States, muses, "You know, the one thing I learned from my patients: they all hate the phone company." Later, when he's been kidnapped by phone company automatons, Coburn is subjected to a corporate-style presentation of a future with wireless telephones implanted in subscribers' brains, illustrated with an animated cartoon of a phone being injected into a customer.

"We call it the cerebrum communicator, or the CC for short," explains the enthusiastic automaton. "This dandy little device can actually perform every function of the old-fashioned telephone and more, and it does it without any costly maintenance. Without telephone poles, without wires, without exchanges, without anything, in fact, except another CC in another location." The automaton has a crazed looked on his face as he describes the new product to Coburn, who's being held captive in a locked telephone booth. "Now, you're probably wondering why we made it so small. Because it will be in, and powered by, your own brain!" Coburn looks in shock as the automaton concludes his sales pitch. "Can you imagine the ease, the fun, with which you could place a call? Why, all you have to do is think the number of the person you wish to speak with and you are in instant communication, anywhere in the world!"

Hank Azaria phoning from the set of The Cradle Will Rock

Wireless phones play critical roles for both the cinematic good guys and bad guys. In the kidnap drama *Ransom* actor Mel Gibson, in a contest against a crazed cop gone bad, is able to call the FBI on his car phone and tip them off as to the identity of the kidnapper. Film star Bruce Willis struggles against narco-terrorists in *Die Hard 2*, where wireless telephony becomes a dangerous tool when commanded by a loose-cannon TV reporter who holes up in an airplane toilet and broadcasts a live report via the airplane phone. And in the President Clinton send-up *Wag the Dog*, wireless telephones keep the action moving.

Real life
The above was all Hollywood, but there are some amazing stories of cell phone use—and misuse—in real life as well.

Radio star Howard Stern is dismissed by some

listeners as nothing more than a dirty-talking prankster. But just before Christmas in 1994, he used his radio show to talk one of his fans out of jumping off a New York City bridge–connected to the would-be suicide by mobile phone.

During the hijacking of a Spanish airliner, passengers used their mobile phones to call police from the Valencia airport tarmac to inform them that the hijacker, despite his claims to the contrary, was unarmed and was carrying no bomb. The hijacker was persuaded to give up when he was connected–via mobile phone–with his psychiatrist, who said later about the call: "After talking with him for four minutes, he softened and the situation was resolved."

These are not isolated mobile-phone-as-hero stories. On the road, problems ranging from traffic jams to disastrous wrecks are routinely reported to authorities by travelers equipped with phones.

Another dramatic example of mobile phone heroism was the self-rescue of a busload of scared passengers in Massachusetts, as reported by the *Cape Cod Times:*

> The trip–described by some of the fifty-four passengers as a nightmare–was on a Plymouth and Brockton bus headed for Park Square, but the bus driver was having a tough time handling his task. Passengers say they saw him dozing off, barely able to keep control of the bus.
>
> "He seemed to lean forward as if he was falling asleep, then he would snap his head

backwards," Kelley Meier of West Yarmouth said.

Meier and about a dozen other motorists called police on their cellular phones because of the bad driving.

"They all reported he was all over the road, driving very erratically," said Sergeant Larry Gillis, a state police spokesman.

The bus driver, Stephen Souza, 31, of Plymouth, reportedly tailgated cars, weaved in and out of traffic, and nearly drove off the highway. At times he drove well below the speed limit. Other times, he traveled at sixty-five mph.

State troopers were waiting for the bus after it got into the Southeast Expressway's enclosed commuter lane in Braintree. Gillis said stopping the bus caused a huge traffic backup, which ended only after a tow-truck driver moved the bus to a turnout in Milton.

Souza was cited with driving to endanger, impeded operation, and marked-lane violations. The passengers had to wait an hour in the turnout while another driver traveled from Plymouth to complete the journey Souza started.

A tear jerking story with a happy ending started with a kidnapping and robbery in Tampa. The bad guy forced a mother to take him to the airport, where he robbed her and locked her in the trunk. But she managed to dial 911 and slip the cell phone to her three-year-old daughter with the hurried instructions: "Keep talking to whoever answers!" The 911 dispatcher instructed the little girl to honk the horn to help the police locate the car. "My daughter's not allowed to honk the horn," mother Mary Graves told *USA Today*. "But when she started, I was screaming, 'Honk it baby, you just keep honking.' I don't think I could have lasted much longer. It was so hot."

Many people buy mobile phones for security. Motorists use them to report accidents and drunk drivers.

In the United Airlines *Hemispheres* magazine, Santiago-based journalist Cheryle Stanton reported that driving while talking on the phone is illegal in Chile. But police there find that when they pull over drivers holding handsets that about a third of the phones are fake, held in full view flaunting the law for their potential prestige only.

Lovers' laments

Stories abound of difficulties in managing all the details of new-found technology. Many a love affair has been found out by spouses stumbling upon messages from illicit lovers on home answering machines because the intended recipients forgot to switch off the call-forwarding feature on their mobile phones. Other mobile users forget that their e-mail or pages might be stored in their handsets if they are not erased, and those stored e-mails can be retrieved, in all their steamy glory, by a suspicious spouse—or just one who is bored and picks up the handset to fool with its intriguing features.

> **MARY GRAVES'S DAUGHTER TALKING WITH THE TAMPA 911 DISPATCHER**
>
> *DISPATCHER:*
> Are you there?
>
> *THE GIRL:*
> Yes. I want my mommy.
>
> *DISPATCHER:*
> Are you in a car?
>
> *THE GIRL:*
> Yes.
>
> *DISPATCHER:*
> Do you see any airplanes?
>
> *THE GIRL:*
> Mommy's in the trunk.

Wireless literature

Contemporary literature is embracing mobile telephones as a plot development tool and a critical component for replicating today's landscape. In Nicholas Evans's bestselling *The Horse Whisperer*, New York magazine editor Annie Graves is on a train when her husband Robert wants to contact her about their daughter Grace's condition. "Then Robert called her on her cellular phone," writes Evans. "He was at the hospital. There was no change." After conversation about their daughter, Robert offers to meet Annie at the train station. She tells him not to meet her, to stay to the hospital with Grace, as she will take a cab. "Okay. I'll call you again if there's news," he says. It is a scene that clearly establishes the story as occurring after the ubiquitous spread of cellular phones, one that couldn't have been crafted before the spread of the technology.

Do You Talk Directly Into the Telephone?

The proper way to use the telephone for best results is to hold the transmitter directly in front of the lips while you are talking.

Do You Speak Pleasantly?

It may be your best friend or best customer. Greet him as pleasantly as you would face to face. Pleasant people get the most fun out of life.

Do You Talk Naturally?

Normal tone of voice is best. Whispered words are indistinct. Shouting distorts the voice and may make it gruff and unpleasant.

Do You Hang Up Gently?

Thoughtlessly slamming the receiver may seem discourteous to the person to whom you have been talking. It's better to hang up gently.

Do You Answer Promptly?

Delay in answering may mean that you miss an important call. The person calling may decide that no one is there and hang up.

W e've all encountered the idiots talking in-
tently on their phones in the freeway fast
lane, going slowly or weaving around because
they're paying attention to the conversation and
not the traffic. Or we've been face-to-face with a
business colleague in an important meeting when
suddenly we're ignored in favor of a ringing phone
in a jacket pocket.

The telephone is a curious instrument because

Woody Allen

it acts as an extension of what
comes so naturally to people: conversation. People love to talk. We
need to talk—whether chatting in
a café over a leisurely cup of coffee, debating in school or at work,
or whispering to a lover. Nothing
is more of a common denomina-
tor among human beings than
conversation; it is one of the traits that so
clearly sets us apart from other animals.

When we pick up a telephone, it is easy to
lose ourselves in the intensity and focus of the one-
to-one conversation that occurs between callers.
Plenty of rude mobile telephone users do not real-
ize they are being impolite. Their world is momen-
tarily restricted to the conversation with the party
at the other end of the connection.

Flesh comes first

The basic rule in polite society is easy and obvious:
when you are interacting with a companion in the
flesh, that in-person relationship always has prior-

ity over a telephone call. Just because a phone is ringing does not mean it must be answered. But if when your phone rings you feel a need to directly check on who is

Warren Miller

"Oops! Sorry, Ned. I guess I dialed you by mistake."

calling, offer a sincere "Excuse me!" to whomever you're with before answering the call. In most cases it is appropriate to remove yourself, at least somewhat, from your in-person companion. Stand up and walk to a corner of the room or out to a hallway, or at least turn your face away. Leaving someone just sitting in front of you while you start and proceed with a phone call almost always leaves the companion you ignore feeling abandoned and reduced to secondary importance.

But remember: as offensive as it may be to talk on the phone in front of your flesh-and-blood companion, leaving someone sitting alone while you excuse yourself for a phone call must be done sparingly. Frequent interruptions favoring your telephone are quite rude.

Exceptions to the basic rule

Of course there are exceptions. With close friends or colleagues you can make it clear that you must take a call for one reason or another, and then even cover the mouthpiece for a second to tell them who is calling and that the call will be brief. If you know that you will probably be receiving a call during a get-together with others, you can inform them ahead of time, apologizing in advance that you'll need to take an incoming call, requesting their understanding.

Quickly taking a call and telling the caller that you're otherwise engaged but will call back is a decent alternative to using your voice mail if you want to make sure you make personal contact with incoming calls. But leaving your first caller in silence while you stay too long on the second call is just as rude as ignoring someone you've been talking with in person because a third party you find more interesting has entered the room.

Restaurants are not phone booths

Taking calls and phone conversations (especially in a loud voice) are particularly offensive in a restaurant dining room. If you choose to accept calls

during a meal, the best alternative is to answer the phone as you're excusing yourself from the table and tell the caller to hang on for a few seconds as you find an appropriate place to talk. Hold your

"Everything's fine, thanks."

conversation outside of the dining room, or even outside of the restaurant on the street. A terrific alternative is a telephone booth. Many restaurants are equipped with phone booths or phone alcoves, and you can take advantage of their privacy while using your own wireless handset.

Some restaurants with cloak rooms or attentive hosts will—if you trust them—take your phone

while you dine, answer your calls, and either take messages or summon you from your table for those calls you'd stated you would take immediately.

"Hey, Frank! I've been on hold with you for the last 20 minutes!"

Other restaurants and bars enforce a strict policy forbidding mobile phone use. Of course, if you use the phone's vibrator instead of the ringer, no one need know you have a call as you excuse yourself and walk outside to answer your phone.

Yet there are restaurants that do not object to mobile phones in their dining rooms and even encourage their use. "We love them," is the response of Colleen McShane, in her role as executive director of the Illinois Restaurant Association. "We think they're absolutely wonderful. They allow people to go to restaurants and still do business, and be able to stay longer."

Pounding the pavement

We can consider streets and sidewalks completely appropriate for walking and talking, although it's important to remember how easy is can be for an intense conversation to distract you from paying attention to traffic.

Nonetheless, it is basically rude to be engaged in a loud and animated conversation that disturbs others, even outdoors. Just as you might not appreciate the music blasting out of some punk's boom box, your yelling into your handset—whether it be abuse to a co-worker, orders to your broker, or passion for a lover—is probably more annoying than amusing to passersby. Given how sensitive the technology now is, you can talk softly and the person on the other end of the connection can understand you just fine.

Phone-free zones

Start important business meetings by announcing that you're turning off the ringer on your phone and that you expect your colleagues to do the same. Again, your voice mail can collect your messages. In

Sign posted on the door of Café Ebeling, Amsterdam

addition, by shutting off your phone's ringer you make it clear that you consider the in-person meeting important enough that it should not be disturbed, and that you expect those meeting with you to reciprocate.

As is the case in a café or restaurant or just a casual meeting with another, the idea is to be in control of the technology, not let the technology control you. You wouldn't hold an important business meeting with the door open, allowing the agenda to be interrupted whenever someone wandered in to bring up matters that had nothing to do with the meeting. Before the advent of wireless technology, we used to advise the secretary, "Hold all calls!" Keeping your mobile phone quiet during a meeting is the logical extension of that common practice.

Of course you send a very obvious message if you choose to take calls during meetings, and if you're the boss—or quite confident in the security and importance of your position—you may decide to err on the side of rudeness.

Lecture halls and classrooms are on the list of those places where you simply should never speak on your wireless phone, but that does not mean you must be out of touch. With the ringer set to vibrate silently you can excuse yourself and take your call in the corridor without your classmates or professors knowing that you're choosing a phone call over the lesson.

Concerts, live performances, and movies are more problematic. Leaving your seat suddenly is inappropriately disruptive; unless you expect an emergency, you should rely on your voice mail and check it after the show. And if you expect an emergency, maybe you should attend the event another night. If you really must stay in touch, you can take advantage of e-mail and voice-mail technology and inform your correspondents to send you messages when you don't answer their calls. Then you can collect your mail through your phone without disturbing those around you.

Correcting the faux pas

If you forget to turn off the ringer and your phone goes off, say, in church or during a play, the best course of action is to grab your phone and switch the ringer off immediately. Do not be tempted to answer the call; your answering machine will take care of it. And if

Henry Martin

"May I call you back? I'm right in the middle of a commencement address."

you do answer it, you'll be exacerbating an already irritating social blunder.

More and more event organizers are taking the initiative and demanding that mobile phone ringers be shut off. Such instructions are appearing in theater and concert programs or are being added to established oral announcements forbidding sound recording and photography. Audiences at school graduations are being asked to switch off their phone ringers before the ceremonies start. Obviously such common sense guidelines need to be followed even if no specific request is made.

"I'm on my cell phone."

Spare the rest of us

Small enclosed public spaces are usually suspect for mobile phone use. A crowded elevator is a good place to simply ask anyone who calls to wait for a moment until you get to your floor, especially if you don't appreciate eavesdroppers. Buses and trains are other places where extended phone conversations can disrupt other passengers' expectations for peace and quiet.

Passengers on most commuter trains around Tokyo may no longer use their mobile phones on board. In Switzerland, the railroads are dealing with increased mobile phone use by setting aside special cars for business travelers who wish to talk on their phones, and we can expect this segregation trend to spread. A survey of British rail travelers found that both leisure and business passengers favor separate cars for mobile phone users. As is the case with restaurants and bars, some trains are equipped with phone booths, which provide privacy while using one's own phone.

Technology can also be used to enforce rules, regulations, and etiquette. Chiltern Railways engineers in the United Kingdom devised phone-proof railroad cars by covering the car windows with a metallic microfilm that the mobile phone radio waves can't penetrate. An Israeli company called Netline Technologies came up with a jamming system for concert halls, theaters, and churches. They call it C-Guard and promise that the low-level radio signal silences mobile phones by blocking radio contact between the handsets and their base stations. Incoming calls cannot connect, so there are no ringing phones in the hall, and outgoing connections are impossible to complete, so no loudmouths can interrupt the proceedings.

Netline Technologies suggests that customized versions of C-Guard could be used to block all mobile phone usage except for those phone numbers authorized to penetrate the jamming. Since the apparatus generating the constant radio signal used to jam phones is so small, the potential for abuse of a product such as C-Guard seems enormous—from individuals who simply don't want to be bothered by mobile phones to nefarious characters intent on disabling mobile communications in order to expedite their questionable activities.

Caller consciousness

The responsibilities for courtesy are not limited to the end user; wireless etiquette extends to those who call mobile phones as well. For example, there are probably inappropriate times to contact a friend or colleague who you know carries a mobile phone, and you should be sensitive to such interruptions. Since the devices can be shut off, and since most

"Nice talking to you, Al!"

are equipped with caller ID features, you might feel comfortable about calling anyone with a wireless whenever you wish. But, as is so often the case, common courtesy is the best guideline for your behavior. You should assume that if a correspondent of yours uses a wireless device, it's because he or she needs to be available at odd hours. So if you call in the middle of the night or at dinnertime, expect to wake or interrupt whomever you call. In general, unless you've been invited otherwise, the best tactic for calling a wireless is the same as calling a more traditional phone: during regular business hours for business calls, and not too late or too early for personal calls.

If you call someone on their mobile phone and reach them in a noisy place where it is difficult for them to hear you, the onus is on you to speak loudly and clearly, or–if the call is not urgent–to ask when it would be convenient to call back once the person you've called has moved to a quieter location. Although late-model mobile phones can transmit voices with quite unbelievable clarity even if there is a lot of background distraction, it still can be difficult to hear an incoming caller if you're standing with your mobile next to a jack-hammer or an idling bus.

The wireless as leash

One of the strange sociological manifestations of mobile telephone technology is that once friends and colleagues know you carry a mobile phone, they *expect* to be able to get in touch with you whenever they wish and often feel insulted if you are not responsive to their rings.

When calling wireless phones it's prudent to take into account that the friend or business associate we're calling might be in the bathroom, driving in dangerous traffic, in a crucial meeting, walking–hands full of groceries–or just ignoring the phone in favor of a gorgeous sunset.

If the person on the other end of the connection is using GSM technology, there is another factor to take into account before you dial: what time it is where that phone is about to ring. GSM allows a phone company subscriber to

> *"I used to date a girl who would get noticeably upset when she couldn't reach me right away on my wireless," is the lament from one frustrated boyfriend. "This was a problem. I did not buy the phone to be on a wireless leash, though she seemed to think of it that way."*

use a mobile phone all over the world without changing phone numbers. That means you could be dialing a number you think is local because it shares your area code, but the phone rings half a world and a dozen time zones away, waking your poor jet-lagged friend from a deep sleep in a faraway hotel. If you do make that mistake one thing is certain to occur next no matter how hard you try to avoid it. After you apologize, you will invariably say, "What time *is* it there?" followed by a query about the weather.

Another courtesy worth extending to those you call on a wireless phone is a brief, not lengthy, conversation. In most parts of the world the calling party pays for any calls to wireless phones. That means that the mobile user is not concerned with

the length of the received call because no extra charges are incurred by answering the mobile phone. But in the United States and Canada, the mobile phone users generally pay for the air time whether they make or receive the call. Consequently, it is decent etiquette to keep your calls brief unless you're told not to worry about the charges.

Is that a phone in your pocket, or . . .

There are a few distinct ringing sounds shared by most mobile telephones. It is natural–especially in a crowded environment–to reach for your phone when you hear a ring that sounds like yours. And often, as you grab at your pocket or purse on the first ring, someone nearby picks up their phone and says, "Hello!" The call wasn't for you after all, and you–correctly–feel silly. Here's the cure: Don't

"Dave, would you hold on a sec while I take care of some personal business?"

jump for your phone until after the first ring. By then you'll usually have had a chance to determine if the ringing is for you. Even if it is, you can answer the call with some grace instead of looking as if you're so tethered to your electronics that they and whoever is calling you are in charge of your life.

Take a moment before grabbing your phone and pushing at the buttons; pushing the wrong buttons can result in disconnecting the incoming call before you get a chance to chat.

Consider how you look to others when you're talking on your phone in public. Sticking a finger in your free ear to improve the acoustics, for example, looks awful, and who wants to shake that hand later if they've been watching?

Saying good-bye

If you're talking on your mobile phone and the battery is about to lose its charge, it is rude to justify ending the conversation

> **THE MOBILE PHONE AS STATUS SYMBOL**
>
> *Tossing your phone on a bar or restaurant table so it is easily available to interrupt your drinks or meal is gauche. Phones are small enough now that there is no excuse for taking them out of your pocket or purse.*

If you're seduced by the idea of calling radio talk shows, you should always inform the call screener at the radio station that you're calling from a mobile phone. The wireless connection continues to hold quite a bit of caché for talk shows and you often are advanced to the front of the waiting line just because you're calling from the road.

because of the lack of electricity. Offer to find a traditional phone, or arrange for a specific time to continue talking, or tell the person you're speaking with that as soon as you find an electrical outlet you'll plug your phone in and call back. If you want to end a conversation, you should end it honestly and politely.

Although wireless communications networks are constantly improving and are often completely clear and reliable, there are times when the signal deteriorates or you move out of range. It's a good idea to tell those you call that you're talking from a mobile phone so that if the connection fades or drops, they'll know to wait for the few seconds it might take for clarity to resume.

Driving and talking

Talking on the phone while driving can be quite distracting. Just as is the case in walking down the street, or sitting at a café, it  is easy to get deeply involved in a phone conversation. These one-to-one connections we make over the telephone are often intense, and we're accustomed to providing our undiverted attention to the telephone call. Obviously such complete focus on a conversation can be disastrous if you're also piloting an automobile through traffic. If you must engage in multi-tasking, the phone call must take second place to driving.

Do your best to use a hands-free telephone in the car. Juggling the phone, the steering wheel, the gearshift, and a cup of coffee while you try to take notes on the conversation is a bad idea. If you must write something down, look for a place to pull over. If

"Wonderful! This way I can go block after block talking to myself and nobody looks at me as if I were crazy."

your conversation is charged with emotion or revolves around a critical business deal, pull over. Pulling over adds the bonus of guaranteeing that you won't drive out of the coverage area and lose the connection.

In some countries it is illegal to drive and talk on the phone; check on the local laws when you rent a car or apply for your visa to go abroad. In some cases existing laws are used by police to cite drivers talking on the telephone. In England, for example, Gloucestershire police set up a checkpoint on the A417 looking for drivers they considered not in proper control of their cars. They stopped drivers not only for talking on the phone but also for looking at a map, and eating a doughnut! Some were warned, others were fined—few were happy.

"We got a pretty negative response," one of

the officers in charge of the operation, Ian Lloyd, confessed to the London *Daily Telegraph*. "They do not appreciate being told that they are not in full control if they are using a hand-held mobile. They all think they are brilliant drivers who are always safe on the road whatever they are doing. The fact is, they do not stop to think what would happen in an emergency. Using a mobile is not an offense in itself, but it can distract a driver's attention."

Of course so can blaring music, a dramatic radio show, or an attractive traveling companion, and–apparently–an appealing doughnut, at least in Gloucestershire. The obvious solution is to make sure you always give priority to your driving when you use a mobile phone on the road, and the best device for minimizing the physical distractions of using the phone is to take advantage of hands-free speaker phones.

If you decide to use the phone while you drive –and most of us do– it's a terrific opportunity for getting work done during commute traffic

Courtesy of Motorola Museum c. 1998 Motorola, Inc.

Paul Galvin, Motorola founder, on car telephone in 1946

and is clearly a convenient use of otherwise down time–consider your route and environment. Rolling up the windows reduces the street noise and makes it easier to hear at both ends of the connection. It's a good idea to turn off the radio. Don't start a call just before probable interference, such as driving through a tunnel.

Headsets are available for car phones that combine an earphone and a microphone, allowing for hands-free operation that reduces road noise for both parties in the conversation. But, just as is the case in listening to music over earphones while driving, using such a system reduces a driver's awareness of crucial traffic sounds.

Stressful calls in the car should be avoided. The middle of rush hour is poor timing for an anxiety-filled telephonic fight with a spouse or negotiation with a boss or client. This is not only a question of safe driving. If your concentration is split between driving and an important call, the call may not be receiving the required level of your attention either.

If you are involved in an accident while talking on the telephone, you can expect that your divided concentration may well be used against you by the police and the courts when they determine who was at fault.

Initial academic studies of driving and talking support the conclusion that it's a potentially dangerous combination—as does simple common sense. Following the publication of a Canadian study that showed drivers using mobile phones were a greater accident risk than those who didn't talk on the phone while driving, the magazine *Wireless Week* checked in with a couple of the largest insurance carriers for their responses. Neither State Farm nor Allstate planned to raise rates for those who use phones in their cars. "We would have to have some means of identifying crashes in which a phone might have played some role," said Dave Hurst at State Farm, adding, "I really can't guess how this might be done. Industrywide statistics on specific claim problems are difficult to come by." At Allstate, Elio Montenegro dismissed the idea of higher rates for phone users with a simple answer: "The way we'd look at cell phones is the same way we'd look at CD players."

But Professor John Volanti at the Rochester (New York) Institute of Technology disagrees. "They kind of forget about the rest of the world," he generalizes about car phone users. "They're not intentionally cutting people off, they're just not seeing them." Professor Volanti studied accidents in

The Bell Atlantic Pioneers, Nova Five Chapter

New York state and determined that drivers with car phones are more likely to crash than those without phones. Similar conclusions were reached by researchers at the University of Toronto. But raw statistics can be misleading. The studies' results do not necessarily mean that there is a causal relationship between car-phone use and wrecks. Nor do they take into account the beneficial role of the car telephone.

It was difficult to drive and talk at the same time with this early experimental AT&T car phone.

The *San Antonio Express-News* found at least one car-pool mom who considers her car phone crucial to safe driving and keeping the kids under control. "Sometimes when a child is misbehaving," Deidre Alterman told the paper, "I threaten to call his mommy. It works!"

"Neither lender nor borrower be..."

Convenient as a mobile phone is, it can be tempting to ask to borrow one. Resist that temptation. Calls can be expensive, depending on the plan and company a customer uses. Don't reach for another's phone unless invited.

"BE CULTURED AND REFINED,
NOT COARSE AND BOISTEROUS,"
SAYS PROFESSOR HILL.

*"Professor" Thomas E. Hill certainly is not
synonymous with etiquette today—he's been
replaced by Amy Vanderbildt, Emily Post,
and Miss Manners. But back in 1873, as
Alexander Graham Bell was perfecting the
first telephone, the professor wrote a bestselling guide called the* Manual of Social
and Business Forms. *We can still look to
the* Manual *today for guidance as we reach
for our mobile telephones.*

*"Do not talk very loud," teaches
the professor. "A firm, clear,
distinct, yet mild, gentle and
musical voice has great power."*

U se your phone, go to jail. That's reality in some cases, and some of the places and circumstances where the mobile phone is absolutely illegal are quite surprising.

Hello, Bambi

Consider deer hunting in Ohio, for example. A law that predates the extensive use of mobile phones forbids the use of electronic voice transmitting devices while "pursuing, shooting, killing, following after or on the trail of, lying in wait for, or wounding wild animals while using a hunting or trapping implement such as a firearm or a bow." Mobile phones fall under the law's restrictions. The idea is to prevent hunters from conspiring with the help of their phones against their prey.

The Ohio Division of Wildlife enforces the law as part of its effort to promote what its literature calls "the understanding and practice of ethics and fair chase while hunting."

The Sierra Club promotes the carrying of mobile phones by backcountry hikers, but advocates their use only in emergencies. "Be sensitive," reads the club's guidelines, "to the fact that you will encounter people in the backcountry who are seeking their own kind of wilderness experience and try not to intrude."

The in-flight explanation

Airline flight attendants make it clear during their preflight announcements that you may not use your mobile phone once the aircraft starts moving.

Many airlines allow the use of phones when the plane is parked, but some zealous stewards and stewardesses even object to that harmless practice. This policy is not an attempt to force passengers to use the high-priced phones in the seat backs of most airliners. FAA regulations forbid the use of personal mobile phones in the air. The stated warning is that the phone signals might interfere with the plane's navigational equipment. In reality, the point is moot. Technically your mobile phone can't operate when high above the base stations it needs to contact for connections with the rest of the world—the base stations aren't designed to coordinate which one should carry your signal when transmitted to them from above.

Laws on the books

Keeping track of the various laws regarding mobile phones is quite a project. In the Australian state of New South Wales, for example, you can talk on a hands-free phone while driving, but you must stop to dial. Any phone use by drivers is banned in Israel and Brazil. In Singapore, drivers who use their phones are not only subject to a ticket and a  fine of just over a hundred dollars, police may confiscate their telephones. In the Philippines, it is illegal to drive and talk on the phone in Manila and Quezon, but apparently there's no restriction in the rest of the country. Poland too only regulates mobile phone use in urban areas. Ohio and some other states toyed with the idea of outlawing phone calls of more than a minute long. An obvious problem: how to enforce such legislation? Car rental companies in California must provide instructions for the safe use of mobile phones when they rent a car equipped with a telephone. Pulling over to make a casual call is a bad idea on California freeways, where non-emergency stops are illegal. Florida law specifically mandates that a driver using a phone must leave one ear free for traffic noise; in Massachusetts one hand must be on the steering wheel during phone calls.

Hospitals are worried about mobile phone interference with equipment, and many hospitals ask visitors both not to make or receive calls and to turn off their phones. The concern is that even an idling handset may interfere with sensitive hospital electronics.

If you choose to violate these dictums, you

could be discovered. A German company, Rodhe & Schwarz Engineering and Sales GmbH, is out with a product they call Mobifinder. The company claims the unit can locate mobile phones within a fifty-meter range and can be used to enforce bans in places such as hospitals.

Those people who use a pacemaker must be concerned about interaction between the pacemaker and some electronic equipment, such as airport metal detectors. Doctors do not routinely advise their patients with pacemakers to avoid using wireless phones. Researchers studying the potential effect of some wireless handsets on some pacemakers, however, recommend keeping phones at least six inches from pacemakers and not placing them over the devices even when the phones are on but idle. That means no phones in breast pockets for those using pacemakers and, to be on the safe side, hold the handset to the ear opposite the pacemaker when using the phone.

When mobile phone calls are made, the equipment registers the caller's whereabouts within the mobile system. In other words, callers leave electronic footprints within the records of the phone company that manages their outgoing call. This can be a valuable tool for law enforcement.

Help!
Of course, in an emergency, all the rules go out the window. Wireless telephones are modern lifesavers, summoning needed help all over the world every day of the year. Tens of thousands of critical calls are made daily just in the United States to 911 operators across the country.

EPILOGUE

As I was finishing *Wireless Etiquette,* one of my colleagues stopped in for dinner at Wolfgang Puck's signature restaurant in Palo Alto, California. After the meal he spotted Puck and asked the Spago owner what he thought of patrons using their wireless telephones during a meal.

"I'm not a policeman," replied the restaurateur. "People just need to use common sense. If you're with your girlfriend, you shouldn't be on the phone the whole time. Otherwise, she might kick you out." He paused and smiled. "Only Warren Beatty can make love and talk on the phone at the same time."

Puck's advice is appropriate, of course. Much of the guidance in this book is common sense; common sense too many people are forgetting. But Wolfgang Puck's quick answer and specific opinion is representative of a phenomenon I discovered with just about every person I spoke with as I was researching and writing this book: wireless phones influence everyone. Everybody has an opinion about them and how they should or should not be used. Like television, automobiles, clothing, and food –wireless telephones now touch all of us.

Two friends of mine make my point about the universal impact of wireless phones on all our lives. One of them is the quintessential model of the modern global citizen who is always connected with everybody. He's a businessman headquartered in Amsterdam with interests that run from radio stations to bakeries, dry cleaners to chocolate factories. I know I can reach him whenever I want to because he always answers his telephone.

"What's all that yelling?" I asked him once when we were about to discuss a pending deal. He was in the stands of a Dutch football stadium, cheering for his home team and talking business with me half a world away.

That he is available to me whenever I call is convenient for me, but when I'm sitting in an office with him, his lack of decent wireless etiquette can be quite disconcerting. His phone rings constantly, and he reaches into his pocket in the middle of my or his sentences and holds multiple cybermeetings while I sit in person and wait. I hope he reads this book thoroughly.

My other example friend is an academician, a fiction writer who studied for his doctorate in the libraries of New York City.

"You're writing about mobile phones?" he asked me. "Mobile phones? Those things are everywhere!" His response was less than approving as he described the scene in New York University's Bobst Library.

"It's almost more obnoxious than skateboarding," he complained about the handsets ringing in the library study areas, about the students who use their phones in the library.

But then he smiled and started a story. "There is this guy I see around Bryant Park. He's a street person and he holds a Danish pastry as if it's a phone. He holds a Danish, walking briskly down the street, holding it up to his ear and talking into it. It's a long-style Danish wrapped in cellophane, and he walks with it in earnest conversation." My writer friend clearly is intrigued with the mobile telephone as a new literary device.

Back in 1916, Carl Sandburg recognized the social impact of the telephone with his poem "Under a Telephone Pole," written in the voice of a telephone wire that reports: "I am a copper wire slung in the air, night and day I keep singing –humming and thrumming." We can be sure contemporary artists continue to work at adding the wireless phone to our cultural heritage, with intriguing inspiration such as the pastry talker in Bryant Park and workaholics like my Amsterdam colleague.

Because of my work as a journalist, mobile telephony is nothing new to me. Back in the late Sixties, as a young radio news reporter, I was assigned what I remember to be a Plymouth Valiant. Under the dashboard was a complex of lights, buttons, and switches encased in hard plastic. Attached to the control panel was a heavy plastic handset equipped with a push-to-talk button. I kept in touch with the assignment desk by picking up the handset, pushing the button, and calling for the mobile operator to place my call to the office. Once connected, our conversations were punctuated with "over" since the technology did not allow both parties to talk at the same time.

Talking on the phone while driving down the freeway was big fun.

Years later as an NBC News correspondent, I worked with the first generation cell phones. We lugged the units around Washington, with their heavy attached batteries, amazed that we could dial the news desk from just about anywhere in the District of Columbia and the quality of the connection was clear enough to broadcast live from the streets.

Today, with my tiny GSM handset, I travel the world for work and pleasure. Friends and colleagues dial my American phone number and catch me just about wherever I might be. From Auckland to Zamora, I bypass hotel switchboards (and their ghastly surcharges) and complicated foreign-language public phone instructions.

Using wireless phones for so many years, and watching others use and abuse their own, allows me to appoint myself arbiter of wireless etiquette. I hope *Wireless Etiquette* helps us all accept and embrace this new technology with a chuckle and with a firm understanding that we must control it, not allow it to control us.

Peter Laufer
San Francisco, 1998

ACKNOWLEDGEMENTS

Plenty of my friends and colleagues laughed at the idea of me dictating etiquette; apparently my reputation is not that of the most polite guy on the block. But, in the spirit of Supreme Court Justice Stewart, I know rudeness when I see it, and I've been lucky to receive help from many of those same friends and colleagues with this book.

I specifically want to thank Salli Miller, Susan Roth, Markos Kounalakis, Michiel Herter, Michael Hassan, Thomas Christensen, Susan Burks, and Robert Roche for their valuable contributions. Diablo Press in Berkeley, California, was kind enough to give permission for use of the engraving of Professor Thomas Hill. Hill's complete book is available from Diablo. Thanks also to the Motorola Museum in Schaumburg, Illinois, for opening up their extensive archives to this project.

CHAPTER 1
Page 2. © 1998 P.C. Vey from cartoonbank.com. All Rights Reserved
Page 3. Western Electric ad courtesy of Lucent Technologies Bell Labs Innovations
Page 5. The New Yorker Collection 1994 Bernard Schoenbaum from cartoonbank.com. All Rights Reserved.
CHAPTER 2
Page 8. "Getting Up A Party" courtesy of Lucent Technologies Bell Labs Innovations
CHAPTER 4
Page 16. © The New Yorker Collection 1994 Bill Woodman from cartoonbank.com. All Rights Reserved.
Page 19. © The New Yorker Collection 1995 Danny Shanahan from cartoonbank.com. All Rights Reserved.
CHAPTER 5
Pages 28 & 29 Bell Telephone ad courtesy of Lucent Technologies Bell Labs Innovations
CHAPTER 6
Page 31. © 1998 Warren Miller from cartoonbank.com. All Rights Reserved.
Page 32. © The New Yorker Collection 1995 Mick Stevens from cartoonbank.com. All Rights Reserved.
Page 33. © 1998 Mick Stevens from cartoonbank.com. All Rights Reserved.
Page 35. © The New Yorker Collection 1994 Henry Martin from cartoonbank.com. All Rights Reserved.
Page 36. © 1998 Frank Cotham from cartoonbank.com. All Rights Reserved.
Page 38. © The New Yorker Collection 1990 Mick Stevens from cartoonbank.com. All Rights Reserved.
Page 41. © The New Yorker Collection 1993 Robert Mankoff from cartoonbank.com. All Rights Reserved.
Page 43. © The New Yorker Collection 1993 Gahan Wilson from cartoonbank.com. All Rights Reserved.

Wireless Etiquette author Peter Laufer brings specific practical expertise to this project. He has used mobile telephones since their early and bulky incarnations in  the mid-1960s. An award-winning journalist, he works and travels extensively in the U.S. and overseas, always equipped with a wireless telephone, traveling and working with people who rely on mobile phone technology.

Laufer is the researcher, writer, and announcer of the Omnipoint Business Minute radio feature, and the author of several books, including *Inside Talk Radio* and *Iron Curtain Rising*. With his wife, Sheila Swan Laufer, he cowrote and photographed *Neon Nevada*, an analysis of the impact of neon signs on Nevada popular culture published by the University of Nevada Press.

In addition to having published this book, Omnipoint operates its own GSM digital wireless telephone network.

"*Wireless Etiquette* is a public service— for our customers, as well as for their friends, families, and co-workers," says Omnipoint President George F. Schmitt. "The mobile phones we sell are now so ubiquitous that I feel we have a responsibility to promote their proper use."

Omnipoint is licensed to cover all major population centers from the Canadian border to Virginia, plus extensive areas in the South and Midwest.

Omnipoint subscribers also can use their service in cities all across the United States and Canada, as well as dozens of other countries around the world—all due to the unique international roaming capabilities of GSM communications technology.

Founded in 1987, Omnipoint became the largest publicly-traded personal communications services provider in the United States. The company's stock is listed on the NASDAQ National Market. Its trading symbol is OMPT.

Those interested in more information can visit the Omnipoint Website (www.omnipoint.com) or call 1-800-BUY-OMNI.

This book was designed by
Chris Slattery Design, San Francisco.
The text was set in Goudy,
a typeface designed by Frederic Goudy in 1915.
The book was printed and bound by
Thomson-Shore, Inc., Dexter, Michigan,
on 60lb. Supple Opaque recycled text stock.
The dust jacket was printed by
Valley Color Graphics, Modesto, California.
The managing editor was Terry Phillips.
The editor was Kirsten Janene-Nelson.